Staged Plasma Gas Compression Guns with Emphasis on Electron Plasma Modes for Facilitating Nuclear Fueling of Remote Relativistic Spacecraft

James M. Essig

Edited By Dr. Ron Turner

August 5, 2021 to October 20, 2022

Acknowledgments

Before proceeding with this paper, I would especially like to acknowledge the work of Dr. Ron Turner, distinguished analyst, Analytic Services Inc. (ANSER), who is with NASA/NIAC, for providing significant editing of this paper and offering other critiques as well. I greatly appreciate the mentoring I received from Dr. Turner as well as his suggestions on how to improve the paper. Dr. Turner also helped me with some of the concepts as he recommended various changes and approaches to conceptualize the propulsion methods and infrastructure considered herein. Thus, his work was instrumental in helping me draft a better paper with improved concepts.

Much gratitude also is given to Chuck Hawks for permitting me to include excerpts from his fascinating literature on the specifications of large naval guns. The information he provided grounds the proposed muzzle velocities and loading on the pellet launch system presented in this paper.

Additionally, I would like to give my brother Mike lots of praise for providing insight that helped me manage the social aspects of writing this paper as well as providing me other practical suggestions.

Herein are proposed novel mechanisms for deploying principally nuclear fission and/or nuclear fusion fuel for remotely located spacecraft. The method of fuel placement includes as a primary option electron gas gun systems that launch pellets to muzzle velocities of Keplerian to mildly relativistic scales. Accordingly, the fuel would be substantially prepositioned to be intercepted by relativistic spacecraft. Emphasis is placed on thermonuclear explosives that would be detonated at optimal locations proximate to a spacecraft pusher plate or charged chute system. The charged chutes would be attached to the spacecraft by springlike or other elastic mechanisms to dampen the electrodynamic shock of the explosions on the spacecraft and chute. In the first baseline scenario, a 10,000-metric-ton craft is accelerated to a velocity of 0.01 c and thus a Lorentz factor of 1.00005 assuming that 0.5 of the pellet explosive energy is converted to spacecraft kinetic energy. Here the author assumes that the explosive energy embodied in the pellets would need to be approximately equivalent to 0.0001 the mass of the spacecraft or one metric ton. Assuming the yield of a nuclear explosive pellet is about 0.0025 of its invariant mass equivalent, the mass of the pellet runway would need to be about 400 metric tons. This paper also considers scenarios for terminal craft velocities of 0.1 c and 0.9 c.

This paper is partitioned into several sections as follows:

Acknowledgments

Introduction and Overview

The Electron Gas Gun

Pellet Runway Characteristics

Three Specific Examples of the Primary Gas Gun System Scenarios

- Class 1 Electron Gas Guns Enabling Mildly Relativistic Spacecraft Velocities of 0.01 c
- Class 2 Electron Gas Guns Enabling Mildly Relativistic Spacecraft Velocities of 0.1 c
- Class 3 Electron Gas Guns Enabling Fast Relativistic Spacecraft Velocities of 0.9 c

Sail characteristics.

Appendix 1: Prior Art in Staged Compression Gas Guns and Large Artillery

Bibliography

Introduction and Overview

The author proposes a theoretical concept validation of a staged compression gun able to accelerate projectiles and spacecraft to Keplerian velocities and even to relativistic velocities.

The fundamental concept is to use a space-based linear accelerator to propel small fission/fusion devices to produce a propellant runway. Prior theoretical art explores the basic concepts of fusion runways. Accordingly, nuclear fusion fuel pellets or other fuel pellets would be distributed along a substantially linear path for interception by a spacecraft. The fuel pellets would then be captured and processed in nuclear fusion reactors or otherwise detonated in proximity to a spacecraft. While various means of distributing fuel pellets for deep space travel have been proposed, this paper proposes novel methods of dispensing the pellets.

1. The proposed pellet launching gun innovates over prior art in that it operates via a staged explosive decompression of electronic gas. The explosive process includes thermal gas depressurization and coulombic intragas repulsion.
2. The gas gun dispenses thermonuclear explosive devices in nominal examples of 40 kilograms mass with a yield of about 0.0025 of its mass converted to energy or 22.5 kilotons yield. This is about twice the yield of the explosive device dropped over Hiroshima, Japan, near the end of World War II. Novel research is proposed to enhance the mass specific yield and lower the critical mass required for thermonuclear explosives.
3. In a nominal consideration, the explosive bomb energy is coupled to the spacecraft via a large pusher plate to absorb the momentum of the x-ray and soft gamma-ray radiation of the blast. The pusher plate can also be magnetically insulated or otherwise electrodynamically insulated to back reflect the hot plasma generated by the nuclear explosives. Additionally, the hot plasma from the nuclear explosions can be back reflected by magnetic plasma bottle sails and/or by other sail-type electrodynamic mirrors.

Note that some sentences or other portions of text in this paper are repeated to some extent for reader clarity. The author believes that this is necessary so as to reduce the need for rereading portions of this paper. A motivation behind these circumstances is to provide the reader with more intuitive and easily understood concepts as well as assuring the reader the correctness of the computations repeated.

Geoffrey Landis has proposed a method of propelling a starship by a mass beam. Gerald Nordley proposed a similar concept [1].

Greg Matloff has proposed an interstellar ramjet mechanism enabling velocities of 0.20 c with a pellet stream having an order of magnitude less invariant mass than a standard fusion rocket. [2].

Adam Crowl and Gerald Nordley proposed another mass beam system. [3].

Each of the above proposals has technical merits, but none of these methods specifically employ an electron gas gun, nor do they focus on nuclear explosives distributed at Keplerian velocities with respect to the background that can be considered the 2.725 K cmbr. Moreover, the above references focus on moderately relativistic travel. However, James Essig provides scenarios for spacecraft Lorentz factors at least as great as about 1,500, which correspond to a Doppler blue-shifted cmbr temperature of roughly the melting point of tantalum-hafnium-carbide solutions. Such super-refractories when polishing and incorporated into elongate aspect ratio spacecraft shields can enable Lorentz factors much greater than 1,500 where the conical surfaces are highly relative and highly emissive. However, said extreme Lorentz factor scenarios will be specifically addressed in a later paper by Essig.

The Electron Gas Gun

The muzzle velocities of the guns can be adjusted so that the velocity, location, and timing of the ship arrival per pellet can be properly calibrated.

Launch rates of the pellets can be adjusted or made adjustable as can the yields of the pellets for optimal drive efficiency and abilities of the spacecraft systems to handle the nuclear blasts.

Exiting spacecraft may initially receive each shot at a rate of 1 per second to 1 every 100 seconds to enable timely acceleration. Once the velocity of the spacecraft was brought up to high enough velocity, the frequency of the shot in the background reference frame may be reduced.

In short, the electron gas compression gun has merit in that it may enable relativistic spacecraft velocities without the need for large mass ratios and is simpler in design than traditional mass drivers.

Taken to extreme circumstances, spacecraft propelled by nuclear pellet streams may obtain velocities very close to the speed of light and extreme Lorentz factors. Tantalum-Hafnium-Carbide solutions in shielding material may, in principle, enable spacecraft to attain Lorentz factors at least as great as around 1,500, at which point the cosmic microwave background radiation will be relativistically blue-shifted and aberrated to an apparent black body temperature near the melting point of said carbide solutions.

Now, consider in first order that quadrupling NASA's gas compression gun length and adding additional stages may enable projectile velocities as great as 55,000 feet per second or about 16,500 meters per second. This latter velocity is more than adequate for Mars orbit insertion and also into the main asteroid belt

Using an electron gas as final stage pressure embodiments, much greater velocities such as very mildly relativistic velocities could be obtained by guns of great length mounted on the Moon or interplanetary solar orbit. Velocities on the order of at least roughly 750,000 meter/sec or 0.239 percent of the speed of light may be obtained. Conceivably, higher velocities could be achieved with the electro-gas discharge. These extreme velocities when attributed to pellets on the scale of 40 kilograms/pellet would find great use in forwardly distributed nuclear fission or nuclear fusion fuel pellet runways to enable starships to reach large Lorentz factors with nearly no onboard nuclear fuels. For a 0.0001 light-year long distributed runway and average spacecraft frame acceleration on the order of 0.5 Earth g's or 4.905 m/s^2, a human-crewed spacecraft could obtain a velocity of about 1 percent of light-speed and a Lorentz factor of 1.00005.

The above values for pellet stream velocities are in first order derived from the approximately 16.5 km/sec enabled by hydrogen compression multiplied by the square root of the ratio of the hydrogen atom mass and the electron mass which is 42.8. This is so because the mass of the hydrogen atom is 1,837.17 times that of the electron. Since Newtonian momentum is proportional to velocity for a given mass, but assuming said square root of 1,837.17, there is said square root scaling of momentum, but also said square root scaling of number of times on average a given electron will impart momentum to a gun barrel's wall per second. Here, consider pressure in the ideal gas law is a function of number of particle collisions per unit area per unit time multiplied by twice the average pre-incident momentum imparted to the barrel walls of the colliding particles while assuming elastic collisions. Thus is obtained the approximately one-fourth of a percentage of the velocity of light for the exiting projectile.

Now, as the electron gas density increases, coulombic pressure will factor in, thereby increasing the gas pressure beyond what it would be for a given average electron velocity. So it is conceivable that an electron gas compression gun of many stages could accelerate a round to almost the speed of light. A gun barrel with a high electric flux topological reflector or insulator will make the electron-based coulombic force more concentrated, thus increasing the potential velocity of the round to values even closer to the speed of light to attain extreme projectile Lorentz factors.

An anticipated method of providing electron gas is a powerful electrical discharge system such as by way of banks of super-capacitors. The charge may originate by powerful electrostatic discharge.

Other gun configurations would simply involve a long muzzle for which an electron discharge is timed to occur just as a round passes the locations of electrostatic discharge.

Now Coulomb's law can be used to approximate the coulombic expansive force acting on the projectile. Coulomb's law is as follows:

$F = \{1/[4\pi\varepsilon_0]\}(q_1q_2)/r^2$ Eq. 1

Here, a force at 4×10^6 N is assumed, which is substantiated later below and refreshed every 20 meters.

As a first order consideration assuming the above driving force, consider the following formula for discharged electric charge and pellet charge under the condition that both charge deposits considered are equal in magnitude and in sign.

$[Fr^2 (4\pi\varepsilon_0)]/q_1 = q_2$ Eq. 2

Here F is the coulomic force, r is the average effective distance between each initial electrical discharge, and the pellet and q_1 and q_2 are the electrostatic discharge and the charge on the pellet. Here, r = 10 meters.

Since it is considered that both charges equal, $q_2 = [Fr^2 (4\pi\varepsilon_0)]^{1/2} = 0.21096$ coulombs.

Note that ε_0 is the electrical permittivity of empty space and is equal to $8.854187817620 \times 10^{-12}$ Fm^{-1}.

However, as long as the product of q_1 and q_2 remains constant, the charge can be apportioned in any ratio desired. For example, increasing one of the charges by a factor of 100 in magnitude will decreasing the other charge by a factor of 100 leaves the product the same.

Supercapacitors can easily store one coulomb of electric charge and may be placed every 20 meters along the gun.

For cases where the coulombic expansive forces match the electron gas thermal-pressure-based propulsive force on average, both force components are equal to 2×10^6 N. As long as the sum of both the coulombic and thermal gas thrust components remain at 4×10^6 N on average, it is possible to choose any values for these two force components.

In cases where the coulombic and electron thermal gas driving force mechanisms are energy-wise of limited efficiency, e, then the sum of these two parameters must be increased by 1/e to compensate. Naturally, gun designers are interested in optimizing efficiency, duty cycle of the gun, and of the capacitors.

Magnetic and/or electric fields emitted by the interior barrel walls can tweak the radial positions of the pellets within the barrels to prevent high velocity pellets from contacting the barrel wall.

In order to supply volumetrically embodied charge within the barrels for each discharge, thin wires may be installed and used to conduct the discharge current. Provided the wires are thin enough and low enough in mass, the plasma within the barrel proximate to each discharge in space and time will be mostly by mass comprised of electrons. Since plasmas are effectively superconducting when of suitable densities, simple ionization of the wires can provide a conductor for which the electrons discharged not only produce a very hot plasma but also a means of conduction of suitably long time periods.

The gun barrel may be cooled by passive thermal radiation via the use of radiator fins and/or an active electrically pumped cooling effluent that is passed through relatively simple radiators. Choosing outer gun surfaces that are both highly reflective and yet highly emissive can assist barrel cooling between shots.

An interesting scenario would include the inner walls of the gun barrel being lined with an electrical field reflector or topological insulator. Accordingly, the acceleration of the pellets along the barrel would be made much more efficient as the repulsive force acting on the pellets would initially be much greater than otherwise. However, as the pellet kinetic energy increases per discharge, it is expected that energy conservative principles would remain in effect, thus modifying the pellet acceleration profile accordingly. Depending on the class or type of topological insulators, perhaps even those yet to be developed, additional energy may be attributed to the pellets beyond that which would be conventionally possible due to energy conservation principles. Such extra energy may still be reconcilable with energy conservation principles by theoretical recourse to reified zero point field energy, energy intake from other spatial-temporal dimensions, or hidden electrodynamic principles.

As of 1999, Z pinches were the most intense laboratory X ray sources producing 1.8 MJ in 5 ns fom a volume 2 mm in diameter and 2 cm tall. Powers in excess of 200 TW have been obtained. [4].

A Z-pinch is a device of small volume for which a very high energy pulse of duration on the order of nanoseconds ionizes thin wires to produce at temperature on the order of a few million Kelvins.

So, the ability to produce electrical power surges and high plasma energies for propelling the pellet along the gas gun certainly exists. Accordingly, the energy delivered to the pellet would need to be about 40 MJ per ten meter length sections of the gun. However, the pulse duration can likely be increased to values several orders of magnitude greater than that of the above referenced z-pinches. Moreover, a reduction in surge power-levels for each ten meter section of barrel length by several orders of magnitude will suffice to propel the pellet along the gun.

Pellet Runway Characteristics

The spacecraft and the exiting fuel pellets would have fine course correction mechanisms and would inter-communicate to enable suitable temporal calibration of the spacecraft arrival in proximity to each pellet. Clock features in pellets can be based on robust quartz oscillations mechanisms since atomic clocks may not survive the launch forces.

For example, the pellets may have a charging feature that enables them to use the Lorentz turning force to adjust their flight paths as well as optional solar sails, small rockets, small explosive thrusting charges, and the like. Additionally, the pellets may have laser sails for which a beam of light from stations near Earth can induce a drive pressure on the pellets and/or spacecraft for finer course correction. The orientation of the sails can be adjusted for tacking style thrust to accelerate the pellets or spacecraft out alignment of the pellets and spacecraft velocity vectors.

The energy of the bombs will be translated to spacecraft momentum through interaction with attached sail(s). Note that the shape and size of the sail(s) can be chosen for specific mission criteria. The shape and size of the sails may also be adjustable in flight as can the spacing between lines, threads, or fibers of netlike sails. Such additional morphological degrees of freedom can enable more appropriate interaction of the sail(s) with the interstellar and intergalactic medium including but not limited to natural variations in (1) plasma distribution by density, charge, and species; (2) neutral gas distribution by density and species; (3) ambient stellar and quasar light distributions according to energy spectrum and power flux density; and (4) interstellar and intergalactic magnetic field intensity and vector field orientation. Such considerations can be important for drag reduction. For certain conditions, intentional increases in drag for rerouting and deceleration may be applicable. Such natural variations can enhance or degrade spacecraft performances including but not limited to propulsion system efficiency.

Now, regarding the subject of sail erosion by exposure to interstellar or intergalactic gas, it must be realized that the kinetic energy of a gas atom traveling at a velocity of 86.7 percent of the speed of light with respect to the sail would be equal to the binding energy of roughly 10 billion atoms within a sail of micron thickness. Thus, 10 billion atoms could be dislodged should all the energy of the gas atom be deposited within the sail. Incident gas atoms having even higher associated gamma factors with respect to the starship sail could potentially knock loose even more atoms. Further discussion of sail erosion is included in the section on sail properties.

In order to archive spacecraft deceleration, electrodynamic breaking mechanisms may be employed.

For present consideration, a mechanism of choice would include one or more linear induction breaking mechanisms. Accordingly, one or more large highly conducting or super-conducting coils would be deployed. The coils would build up large currents, which would then interact with the background magnetic fields to bring the spacecraft velocity down to Keplerian velocities; upon which small scale velocity correction rockets would be deployed. These rockets may include nuclear electric systems such as traditional ion thrusters or engines such as Ad Astra Rocket Company's VASIMIR Engine. See references for the VASIMIR Engine.

There is also an option for which a pre-deployed pellet stream displaced in the outer reaches of the solar system would be used to slow the craft. The breaking thermodynamics would be simply the reverse of the positive acceleration process.

To travel back to Earth, another pellet stream would be deployed to accelerate the spacecraft on a heading back to Earth. The thermodynamics of the return accelerating process would closely match that of the initial outbound acceleration.

Once again, yet another pellet stream would be used to slow the spacecraft before obtaining close proximity to Earth. The reason for the latter option is that nuclear explosions near the Earth may result in EMP effects, which can damage satellites and electrical power grids on Earth.

The above proposed propulsion missions can be repeated many times and offer strong solutions to the problems of what to do in the near term with excess fissile plutonium and uranium stock piles

Once space-based mining of minerals becomes possible, the ability to find and refine uranium and thorium ores on the Moon, Mars, Mercury, the asteroids, and the Kuiper Belt and Oort cloud object should enable enough fissile fuels or breeder reactor feedstocks to enable far more nuclear explosive charges to be produced than would be necessary for one or several missions.

In order to fabricate the number of needed nuclear charges, extremely low critical mass fissile fuels may be used for the nuclear primaries along with neutron reflectors to further reduce the critical mass of explosively compressed or combined fissile fuels during the detonation process.

The muzzle velocities of the guns can be adjusted so that the velocity, location, and timing of the ship arrival per pellet can be properly calibrated.

An interesting optimization would involve pellets fired in a direction tangential to Earth's orbit around the Sun and for pellets fired from Earth's orbit, tangential to Earth's orbit as well. This way, these two increased velocity components can enable reduced muzzle velocity.

Pellets may also be fired in such a flight pattern that they receive gravity assists from the gas giant planets of Jupiter and/or Saturn. Accordingly, the muzzle velocity of the pellets can be further reduced. These enhanced efficiencies can apply to many scenarios including but not limited to the ones presented in this paper.

Launch rates of the pellets can be adjusted or made adjustable as can the yields of the pellets for optimal drive efficiency and abilities of the spacecraft systems to handle the nuclear blasts.

Additionally, optimized blends of fissile fuels can be used in the composition of the primaries, thus offering research options to greatly reduce the required critical mass of the needed fissile fuels.

Additionally, U-238 based shots might replace conventional nuclear explosives for which neutron irradiation of the U-238 by a neutron beam originating from the ship would induce nuclear fission of the U-238 shots. The fissioning U-238 may set of a nuclear fusion fuel secondary, which may have a yield 10 to 100 times that of the primary.

These proposed methods entail a long-term commitment and the development of suitable space travel industries. Once enabled, very great spacecraft Lorentz factors can be obtained with very small mass ratios. Pellet launch rates and velocities can be up or down adjusted as needed. The above proposal is bold and costly; however, it is never too early to plan such missions. Once space-based mining and manufacturing develop, the feasibility will greatly increase.

Three Specific Examples of the Primary Gas Gun System Scenarios

Class 1 Electron Gas Guns Enabling Mildly Relativistic Spacecraft Velocities of 0.01 c

Here, the author presents a near term realizable scenario for which a staged compression electron gas gun enables spacecraft velocities of 0.01 c. The author reiterates some of the material in the previous section regarding the 0.01 c scenario just for clarification. This is done so that further elaboration on the 0.01 c case is easily interpretable in context.

In order to accelerate a 10,000-metric-ton craft to a Lorentz factor of 1.00005 and assuming that 0.5 of the pellet explosive energy is converted to spacecraft kinetic energy, the explosive energy embodied in the pellets would need to be approximately equivalent to 0.0001 the mass of the spacecraft or one metric ton. Assuming the yield of a nuclear explosive pellet is about 0.0025 of its invariant mass equivalent, the mass of the pellet runway would need to be about 400 metric tons.

A method of converting pellet explosive energy to spacecraft kinetic energy is the deployment of a large charge sail chute that would be pushed ahead of the spacecraft but which would be attached to the spacecraft by an elastic spring system. The recoiling chute would be pulled back to the spacecraft whereupon the next pellet would detonate.

For explosive charges each having a mass of 40 kilograms, 10,000 devices would be needed. Since each device would be standardized, assuming a unit can be mass-produced at a cost of $100,000, the total cost of the fuel pellets would be $1 billion.

For the anticipated 0.0001 light-year long pellet stream, each pellet would need to be separated on average by 10^{-8} light-years or by 100 million meters.

Assuming gun muzzle velocity of 100,000 m/s, the average time between pellet launches would be 1,000 seconds.

The time averaged thermal loading on the gun would be equal to that of the previous example.

The total time averaged power required to accelerate the pellets would be equal to that of the previous example.

The required time averaged electrical power required to operate the gun would be 200 megawatts over the 1,000 second cycle period.

Assuming an orbital location of one AU from the sun, and an adjustable reflector that reflects about one-half of the incident sunlight but transmits the rest, the author obtains the following relation.

$(4 \times 10^6$ N) = <S>/c = <S>/$(3 \times 10^8$ m/s). So the required sunlight power to counteract the acceleration on the gun during the firing time would be 1.2×10^{15} watts. In this paper, <S> is the time averaged light pressure, and c is the velocity of light in a vacuum.

Assuming a barrel system mass of 5,000,000 kg, the net barrel acceleration during the firing operation would be F/m = [$(4 \times 10^6$ N)/(5,000,000 kg)] = 0.8 m/s^2, and the displacement component of the barrel during the firing process would be x = ½ **a**t^2 = 0.4 m. Here **a** is acceleration, and t is time.

Here are considered gun placements in solar orbit but with adjustable orientation via retro-rockets, solar sails, and/or plasma sails. The orientation of the gun and its orbital radius can be readjusted as needed by sail tacking processes.

For a spacecraft having reached a velocity of 0.01 c such as at the end of the pellet runway, the spacing between the pellets would be about 10 times that of the case where the spacecraft still accelerating was only at a velocity of 0.001 c or about 300 km/second. For the case where the spacecraft started at a velocity of 150 km/second at the beginning of the pellet stream, the spacing of the pellets at the beginning trailing edge of the deployed runway would be about 1/60 times that provided at the leading end of the pellet stream.

Now consider the series summation, $(1 + 2 + 3 + ... + 60)$. The average of this series is about equal to 30. If a mythical computer were to add the first Aleph 0 number of integers and take the average, the average would be one-half of the highest integer. Aleph 0 is the number of positive integer. So in the limit that the number of summed integers, which are consecutive, approaches infinity, the average is equal to ½ the largest integer in the series.

So the two pellets farthest away from the Sun will be 200,000 kilometers apart. The two pellets closest to the Sun will be 3,333 km apart.

For a 0.0001 light-year long distributed runway and average spacecraft frame acceleration on the order of 0.5 Earth g's or 4.905 m/s^2, a human-crewed spacecraft could obtain a velocity of about 1 percent of light-speed and a Lorentz factor of 1.00005.

Several mechanisms might be used to reduce the effective G-forces felt by the crew. Among these is the enclosure of crew members' bodies in hydrostatically sealed breathable oxygenated liquid-containing vessels. Alternatively, perhaps nano-technology types of pressure suits could completely encase the crew members' bodies. The pressure suits might optionally pump high pressure air into the lungs of the crew members wearing them and gradually relax the lung pressure as the rate of acceleration was reduced to more manageable levels. Intense electromagnet fields could be used to induce dipole moments in the atoms and molecules composing the crew members' bodies whereupon the magnetic field(s) would then partially "levitate" the crew members, thereby effectively cancelling undue G-forces. Another option is to use nanotechnology or other precise mechanisms for the installation of an electrical charge within the crew members' bodies where an external electric field would pull on or repel the crew members' bodies, thereby cancelling out excessive G-forces.

Note that the hydrostatic "G-suit" mechanisms may be easier to design and operate. The hydrostatic mechanism with all compressible air or gas removed would be indistinguishable from equivalent pressures experienced by scuba divers that have obtained oceanic depths of over one hundred meters. Assuming the average scuba diver has effective body surface area of 1.5 m^2, a depth of 100 meters places well-balanced compressive force on the body of 150 metric tons force. If it is assumed in first order that a given robust diver has a weight of 100 kilograms, the force on the diver's body is 1,500 times the diver's weight on Earth's surface. A hydrostatic chamber would have similar performances to scuba diver scenarios such as that just presented. So such a system may in first order enable crew members to survive 1,500 Earth g accelerations.

Now it is prudent to address the tissue density variations, especially within the human brain, which is a delicate organ. However, the actual density of the human brain is fairly constant, thus reducing the need for correction from buoyant forces. Where buoyant forces could cause brain injury, the selective electrostatic and magnetostatic magnetization of the human body, especially the brain, can likely enable reduction in intrabody buoyant forces.

Other mechanisms for enabling extremely great accelerations involve the freezing of human bodies to near absolute zero. At such temperatures, water literally becomes harder than steel at STP conditions, thus enabling the frozen bodies to undergo extreme accelerations. When co-application of selective magnetic and/or electric fields are included, frozen but viable human bodies may endure much greater acceleration.

Additionally, placing the frozen human bodies in a protective suit while immersing the bodies in cryogenic liquid helium or liquid hydrogen can enable additional benefit afforded by the hydrostatic G-suit mechanisms presented above.

Ideally, artificial gravitational fields of positive or negative charge signs would be developed. However, there is no substantive evidence that research into such mechanisms has ever yielded observable results.

Note that it is assumed that half of the pellet energy is converted to spacecraft kinetic energy on average.

At a dispensing velocity of 100 km/second, one pellet dispensed per 1,000 seconds on average, and 10,000 pellets dispensed, the pellet runway would take a 10 million seconds to distribute or about 1/3 years. During this time, the pellet furthest from the gun will be 1 billion kilometers distant or about 0.0001 light-years distant.

The United States in reality can support the cost of the above projects since the country was able to fund the war efforts in the Middle East and central Asia during the first two decades of the twenty-first century with a total cost of several trillion dollars.

A spacecraft may initially undergo an Oberth maneuver near the Sun to obtain an existing velocity of 150 km second. The required change in spacecraft velocity under the condition that the O'berth maneuver is not utilized for enhanced velocity is 50 km/second. This velocity is consistent with what can be obtained with solar thermal rockets and nuclear thermal rockets with large mass ratios. The invariant mass of the craft after the fuel has been spent will be the 5,000 metric tons originally conceived.

Δv km/s	{1 + [(2Vesc)/(Δv)]} EXP (1/2)	Exiting Velocity km/s
50	3	150
100	2.236067977	223.6067977
150	1.914854216	287.2281323
200	1.732050808	346.4101615
250	1.61245155	403.1128874
200	1.732050808	346.4101615
350	1.463850109	512.3475383
200	1.732050808	346.4101615
450	1.374368542	618.4658438
200	1.732050808	346.4101615
550	1.314257481	722.8416147
600	1.290994449	774.5966692
650	1.270977819	826.1355821
700	1.253566341	877.4964387
750	1.238278375	928.7087811
800	1.224744871	979.7958971
850	1.212678125	1030.776406
900	1.201850425	1081.665383
950	1.192079121	1132.475165
1000	1.183215957	1183.215957

Table 1 shows incremental velocities for various values of Δv at escape velocity of 200 km/sec.

The spacing between the pellets is best proportional to the projected velocity of the spacecraft at a given pellet-specific location along the acceleration path of the spacecraft in pellet drive mode.

Class 2 Electron Gas Guns Enabling Mildly Relativistic Spacecraft Velocities of 0.1 c

Herein is considered gas guns that may enable spacecraft velocities of about 0.1 c and Lorentz factors of about 1.005. Such velocities can enable timely generation ship travel from star to star for alive and awake crew members and passengers.

In order to accelerate a 10,000-metric-ton craft to a Lorentz factor of 1.005 and assuming that 0.05 of the pellet explosive energy is converted to spacecraft kinetic energy, the explosive energy embodied in the pellets would need to be approximately equivalent to 0.1 the mass of the spacecraft or 1,000 metric tons. Assuming the yield of a nuclear explosive pellet is about 0.0025 of its invariant mass equivalent, the mass of the pellet runway would need to be about 400,000 metric tons.

For explosive charges each having a mass of 40 kilograms, 10 million devices would be needed. Since each device would be standardized, assuming a unit can be mass-produced at a cost of $100,000, the total cost of the fuel pellets would be $1 Trillion.

For the anticipated 0.01 light-year long pellet stream, each pellet would need to be separated on average by 10^{-9} light-years or by 10 million meters.

Assuming gun muzzle velocity of 100,000 m/s, the average time between pellet launches would be 100 seconds. So the pellet runway lay down time would be 1 billion seconds.

Assuming an orbital location of one AU from the sun, and an adjustable reflector that reflects about one-half of the incident sunlight but transmits the rest, the author obtains the following relation:

$(4 \times 10^6 \text{ N}) = \text{<S>}/c = \text{<S>}/(3 \times 10^8 \text{ m/s})$. So the required sunlight power to counteract the acceleration on the gun during the firing time would be 1.2×10^{15} watts. By firing time, what is meant is the time interval that the round travels down the gun barrel.

Assuming a barrel system mass of 5,000,000 kg, the net barrel acceleration during the firing operation would be $F/m = [(4 \times 10^6 \text{ N})/(5,000,000 \text{ kg})] = 0.8 \text{ m/s}^2$, and the displacement component of the barrel during the firing process would be $x = \frac{1}{2} at^2 = 0.4$ m.

Assuming a reverse acceleration time of 100 seconds, the barrel reverse acceleration need only be 8×10^{-3} m/s^2, and thus, the required sunlight power on a sail to counteract the firing acceleration effects is 1.2×10^{13} watts. The latter value can be obtained by a light-sail having collection area of a mere 12,000 km^2. This is a mere equivalent of 109.54 km by 109.54 km. For a sail having capture area of 12,000 km^2 and a mass specific area of one milligram per square meter, the total mass of the sail would be a mere 12,000 kilograms. A sail having area mass density of 0.01 gram per square meter would have mass of 120 metric tons.

The time averaged thermal loading on the gun would be 10 times that of the previous example.

The total energy required to accelerate the pellets would be 1,000 times as great as that of the previous example.

Additionally, the total cost to assemble and launch the pellets would be about 1,000 times greater than in the previous example.

For a spacecraft having reached a velocity of 0.1 c such as at the end of the pellet runway, the spacing between the pellets would be about 10 times that of the case where the spacecraft still accelerating was only at a velocity of 0.01 c or about 3,000 km/second. For the case where the spacecraft started at a velocity of 150 km second at the beginning of the pellet stream, the spacing of the pellets at the beginning trailing edge of the deployed runway would be about 1/600 times that provided at the leading end of the pellet stream.

Now consider the series summation $(1 + 2 + 3 + \ldots + 600)$. The average of this series is about equal to 300.

So the two pellets farthest away from the Sun will be 2,000,000 kilometers apart. The two pellets closest to the Sun will be 3,333 km apart.

For a 0.01 light-year long distributed runway and average spacecraft frame acceleration on the order of 0.5 Earth g's or 4.905 m/s^2, a human-crewed spacecraft could obtain a velocity of about 10 percent of light-speed and a Lorentz factor of 1.005.

The assumption of 0.05 of the pellet explosive energy is converted to spacecraft kinetic energy may actually be quite low. Since the Project Orion craft anticipated shaped charging of the nuclear bombs propelling the spacecraft, which would direct 85 percent of the explosive flux onto the pusher plate, and since 90 percent of the bomb energy would range from soft x-ray to gamma ray radiation, it can assumed that about 76.5 percent of the bombs' energies would be transferred to the pusher plate and converted to spacecraft kinetic energy in first order assuming all photon energy absorbed by the plate would be converted to kinetic energy. Assuming loss of about one-third of the energy due to inefficiencies, obtainable is a good estimate of the bomb energies converted to spacecraft kinetic energy at about 50 percent.

In order to accelerate a 10,000-metric-ton craft to a Lorentz factor of 1.005 and assuming that 0.5 of the pellet explosive energy is converted to spacecraft kinetic energy, the explosive energy embodied in the pellets would need to be approximately equivalent to 0.01 the mass of the spacecraft or 100 metric tons. Assuming the yield of a nuclear explosive pellet is about 0.0025 of its invariant mass equivalent, the mass of the pellet runway would need to be about 40,000 metric tons.

For explosive charges each having a mass of 40 kilograms, 1 million devices would be needed. Since each device would be standardized, assuming a unit can be mass-produced at a cost of $100,000, the total cost of the fuel pellets would be $100 billion.

For the anticipated 0.001 light-year long pellet stream, each pellet would need to be separated on average by 10^{-9} light-years or by 10 million meters.

Assuming gun muzzle velocity of 100,000 m/s, the average time between pellet launches would be 100 seconds.

Assuming an orbital location of one AU from the sun, and an adjustable reflector that reflects about one-half of the incident sunlight but transmits the rest, the author obtains the following relation:

$(4 \times 10^6$ N$)$ = <S>/c = <S>/$(3 \times 10^8$ m/s$)$. So the required sunlight power to counteract the acceleration on the gun during the firing time would be 1.2×10^{15} watts. By firing time, what is meant is the time interval that the round travels down the gun barrel.

Again, assuming a barrel system mass of 5,000,000 kg, the net barrel acceleration during the firing operation would be F/m = $[(4 \times 10^6$ N$)/(5,000,000$ kg$)]$ = 0.8 m/s^2, and the displacement component of the barrel during the firing process would be x = ½ at^2 = 0.4 m.

Assuming a reverse acceleration time of 100 seconds, the barrel reverse acceleration need only be 8×10^{-3} m/s^2, and thus, the required sunlight power on a sail to counteract the firing acceleration effects is 1.2×10^{13} watts. The latter value can be obtained by a light-sail having collection area of a mere 12,000 km^2. This is a mere equivalent of 109.54 km by 109.54 km. For a sail having capture area of 12,000 km^2 and a mass specific area of one milligram per square meter, the total mass of the sail would be a mere 12,000 kilograms. A sail having area mass density of 0.01 gram per square meter would have mass of 120 metric tons.

For a spacecraft having reached a velocity of 0.1 c such as at the end of the pellet runway, the spacing between the pellets would be about 10 times that of the case where the spacecraft still accelerating was only at a velocity of 0.01 c or about 3,000 km/second. For the case where the spacecraft started at a velocity of 150 km second at the beginning of the pellet stream, the spacing of the pellets at the beginning trailing edge of the deployed runway would be about 1/600 times that provided at the leading end of the pellet stream.

Now again consider the series summation (1 + 2 + 3 + ... + 600). The average of this series is about equal to 300.

So the two pellets farthest away from the Sun will be 2,000,000 kilometers apart. The two pellets closest to the Sun will be 3,333 km apart.

For a 0.01 light-year long distributed runway and average spacecraft frame acceleration on the order of 0.5 Earth g's or 4.905 m/s^2, a human-crewed spacecraft could obtain a velocity of about 10 percent of light-speed and a Lorentz factor of 1.005.

Again, in order to produce an average acceleration of 4.905 m/s^2, the pusher plate can be mounted on a spring system that slides along rails for about 66.7 seconds. The pusher plate would be accelerated to 327.16 meters per second and slide along a magnetic repulsive rail to deliver the momentum. The rail would be 10,910 meters in length. Perhaps carbonaceous super-materials can support a rail via sufficient resistance to bending moments yet still be fairly low in mass. Additionally, the pusher plate may optionally but not necessarily be electrically and/or magnetically coupled to the rest of the spacecraft but also have elastic metallic springs attached so that it is not dislodged from the craft.

Alternatively, the peak g-loading can be 49.05 m/s^2, and the pusher plate could slide along a rail for 6.67 seconds. The length of the rail this time will be only 1,091 meters.

Alternatively, the peak g-loading can be 490.5 m/s^2, and the pusher plate could slide along a rail for 0.667 seconds. The length of the rail this time will be only 109.1 meters.

Using shaped charging of the nuclear explosives assuming 0.0025 of the mass of the devices is converted to usable energy and half the blast energy is converted to spacecraft kinetic energy, the author obtained terminal velocities, which are a bit better than would be obtained using nuclear devices of the same invariant mass specific yield but which are carried along in equal mass quantities in the spacecraft instead of in runway pellet streams.

Specifically, for a spacecraft with a nuclear-bomb-based mass ratio of 5 where the fuel mass is four times that of the final payload, the terminal velocity is 0.080273055 c, and gamma is 1.003237537.

Additionally, the stress on the spacecraft is substantially reduced when the devices do not need to be carried along with the craft.

Another merit is that there is no need to shield the devices from otherwise intra-fuel supply irradiation, which can become problematic when large numbers of devices are stored in a compact manner. The need to shield the devices will require rather large additional mass.

However, if it is desirable to carry a spacecraft mass ratio to slow the spacecraft down by reverse bomb blast drive, then accelerating the spacecraft by pellet runway streams makes a lot of sense. Otherwise, the mass ratio will need to be squared in quantity.

Class 3 Electron Gas Guns Enabling Fast Relativistic Spacecraft Velocities of 0.9 c

Herein is considered guns that may enable spacecraft velocities of about 0.9 c and Lorentz factors of about 2.3. Such velocities can enable timely generation ship travel from star to star for alive and awake crew members and passengers.

The really strong benefits are going to come for the 0.9 c case. Since about 90 percent of yield energy of the shaped charged nuclear explosives is in the soft x-ray to gamma ray range energy and about 85 percent of the yield energy can be directed toward pusher plates in a fairly narrow cone, what the author has determined is that the mass of the pellets required to achieve 0.9 c is several orders of magnitude less than the invariant mass of the fuel bombs that would otherwise need to be carried along by the craft. Of course, this only applies to nuclear explosions in space. In the Earth's lower atmosphere, the initial x-ray and gamma ray flash is absorbed by the surrounding atmosphere and other objects in a manner such that about 40 to 45 percent of the yield is light and roughly the same is mechanical blast energy. The other 10 to 20 percent is prompt gamma ray radiation, neutron, fission fragment, and fusion products and residual radioactive fallout energy.

Assuming all focused beam energy can be intercepted by the craft, the only main efficiency losses that will incur will be due to non-ideal alignment of the photon flux with the velocity vector of the spacecraft due to relativistic aberration of the bomb light. The author has developed numerous formulas for these scenarios.

Relativistic red-shifting of the light source will be no lossier than if a coherent single frequency source was used to push the craft forward.

The reason for this is that all the light energy in the conical beam will be assumed to be intercepted by the spacecraft eventually. This is not the case when considering light-sails that are pushed from behind by ambient cosmic microwave background radiation, which is black body in nature. In the case of the CMBR, the wavelengths of the photons are increased in proportion to gamma so that not only are there two-dimensional reduction in the amount of photons impinging on the sail, which would reduce power in ways that scale inversely with the square of gamma, but also manifests an increase of time of arrival for a given number of photons of a given initial frequency that scales with gamma. Since the individual photons are also reduced in energy relative to the space craft in a manner that scales inversely with gamma, there is also a reduction in drive power that scales with the inverse of the fourth power of gamma. Another way of looking at the CMBR issue is that the temperature of the CMBR from in back of the spacecraft reduces inversely with gamma. The power radiated from a black body source of constant surface area, emissivity of one, and temperature T, scales with the fourth power of temperature.

Now desired is an average acceleration of 4.505 m/s^2 in the spacecraft reference frame.

Also desired is the final Lorentz factor of the spacecraft of 2.3.

Now,

F=ma.

So the force acting on a spacecraft having mass of 10,000,000 kilograms for which the acceleration is 4.505 m/s^2 in the spacecraft reference frame is 45,050,000 N.

Now, work = $\int$F• dx. So the total desired total work is $(1.2941)(10,000,000)[9 \times 10^{16}]$ J = 1.1646×10^{24} J.

Now, since F is constant in the spacecraft reference frame, simply divide the work F to get the path length of required travel.

So distance of travel = d = $[1.1646 \times 10^{24}$ J$]/(45,050,000$ N$)$ = 2.58512×10^{16} m.

So assuming that 0.0025 of the fusion fuel pellets is converted into energy and half of this energy is converted to spacecraft kinetic energy, the invariant mass of the pellet stream will be $\{[1.1646 \times 10^{24}](400)(2)/[9 \times 10^{16}]\}$ kg = 1.0352×10^{10} kg.

Since each pellet has invariant mass of 40 kg, there are needed 2.588×10^{8} pellets.

The average distance between the pellets will be 9.98887×10^{7} m = 99,888.7 km.

Assuming gun muzzle velocity of 100,000 m/s, the average time between pellet launches would be 998.887 seconds.

Assuming an orbital location of 1 AU from the sun, and an adjustable reflector that reflects about one-half of the incident sunlight but transmits the rest, the following relation is obtained.

$(4 \times 10^{6}$ N) = <S>/c = <S>/$(3 \times 10^{8}$ m/s). So the required sunlight power to counteract the acceleration on the gun during the firing time would be 1.2×10^{15} watts. By firing time, what is meant is the time interval that the round travels down the gun barrel.

Again, assuming a barrel system mass of 5,000,000 kg, the net barrel acceleration during the firing operation would be F/m = $[(4 \times 10^{6}$ N)/(5,000,000 kg)] = 0.8 m/s^2 and the displacement component of the barrel during the firing process would be x = ½ at^2 = 0.4 m.

Assuming a reverse acceleration time of 998.887 seconds, the barrel reverse acceleration need only be 0.00080099 m/s^2 and thus the required sunlight power on a sail to counteract the firing acceleration effects is 1.201337×10^{12} watts. The latter value can be obtained by a light-sail having collection area of 882.686 km^2. This is a mere equivalent of 29.71 km by 29.71 km. For a sail having capture area of 882.686 km^2 and a mass specific area of one milligram per square meter, the total mass of the sail would be a mere 882.686 kilograms. For a sail having capture area of 882.686 km^2 and a mass specific area of 100 milligrams per square meter, the total mass of the sail would be a mere 88,268.6 kilograms or 88.2686 metric tons.

Now assume that the spacecraft is pre-accelerated to 150 km/s as it becomes coincident with the first pellet, which is assumed to be traveling at 150 km second and that half of the pellet energy is converted to spacecraft kinetic energy. The yield energy of the pellet is $(0.1)[9 \times 10^{16}]$ joules so 4.5×10^{15} joules is transferred to the spacecraft.

So, KE = ½ m v^2. Thus, the velocity increase of the spacecraft will be Δv = $[(2)(KE)/m]^{1/2}$ = 30 km/s.

So the spacecraft after becoming coincident with the first pellet will require either extreme acceleration handling ability or a long pusher plate mounted on a mechanical and/or field effect spring mechanism. A good way to provide a spring mechanism would be to include a magnetically and/or electrically suspended rod mechanism and a pusher plate that would be pushed along the rod against a magnetic bearing located on the stern of the main portion of the spacecraft.

Now solve the following equation for launch time between last two pellets to be launched or the first to be intercepted by the ship. $[4.505 \ m/s^2](t) = 30,000$ m/s. Thus, t = 6,659.2 second. For a = 45.05 m/s^2, 450.5 m/s^2, 4, 505 m/s^2, the launch times are reduced by 0.1, 0.01, and 0.001 respectively. So the distance between the last two pellets to be launched is ½ at^2 to yield 99,886.9 km, 9,988.69 km, 998.869 km, 99.8869 km, for accelerations of 4.505 m/s^2, 45.05 m/s^2, 450.5 m/s^2, and 4,505 m/s^2 respectively.

For the first two pellets to be launched, assuming acceleration of 4.505 m/s^2, the distance between the pellets in the background will be about (99,886.9 km) [(0.9 c)(2.2941)]/[(0.0001 c)(1)] = 2.06235×10^{9} km and the firing time between the pellets is 2.06235×10^{7} s. The spacecraft will experience the time of travel between the pellets in the spacecraft frame as 3,329.55.

For all specific cases considered, that is, the 0.01 c, the two 0.1 c, and the 0.9 c cases considered, a much-reduced mass of each pellet can enable smoother acceleration profiles

By reducing the pellet mass to 4 kg while keeping the invariant mass specific yields and fraction of pellet energy converted to spacecraft energy, the craft will need 10 times as many pellets. However, the acceleration of the spacecraft can be much smoother without the need of such lengthy drive plates.

Reducing the pellet mass to 0.4 kg all else remaining the same, the pellet numbers increase by another factor of ten thus additionally smoothing out the acceleration profile and enabling still shorter drive plate motions and time intervals in the spacecraft reference frames.

By increasing the number of 40 kg projectiles anywhere from 10 to 1,000 fold, while reducing the efficiency of the kinetic energy acquisition 10 to 1,000 fold by the ship, respectively, the spacecraft acceleration can be greatly smoothed out and then so without the need for really lengthy acceleration towers.

In order to produce an average acceleration of 4.905 m/s^2, the pusher plate can be mounted on a spring system that slides along rails for about 66.7 seconds. The pusher plate would be accelerated to 327.16 meters per second and slide along a magnetic repulsive rail to deliver the momentum. The rail would be 10,910 meters in length. Perhaps carbonaceous super-materials can support a rail via sufficient resistance to bending moments yet still be fairly low in mass. Additionally, the pusher plate may optionally but not necessarily be electrically and/or magnetically coupled to the rest of the spacecraft but also have elastic metallic springs attached so that it is not dislodged from the craft.

Alternatively, the peak g-loading can be 49.05 m/s^2, and the pusher plate could slide along a rail for 6.67 seconds. The length of the rail this time will be only 1,091 meters.

Alternatively, the peak g-loading can be 490.5 m/s^2, and the pusher plate could slide along a rail for 0.667 seconds. The length of the rail this time will be only 109.1 meters.

For cases where the pellets are mass-produced, it is conceivable that the cost per pellet might even be reduced to $1,000 at current value if the pellet manufacturing mechanisms can be completely automated by numerical artificial intelligence systems.

Sail Characteristics

The diametrical cross-sectional area of our observable universe is close to 1047 square kilometers and the total mass energy of the observable universe is equal to about 1050 metric tons of which only 4 percent is baryonic. Thus, an average column spanning the diameter of the entire visible universe would have an H2O STP density matter thickness of only about 25 micrometers for reactive matter. However, this is not a concern for the following reasons.

First, the sails could be replaceable grid sails and driven by optical, IR, microwave or radio-frequency radiation. The mass of such sails can be reduced by many orders of magnitude relative to monolithic sails that are only micrometer scales in thickness.

Second, sails having a very thick cable or thread like construction are conceivable where the cables or wires would be many times if not several orders of magnitude thicker than 25 microns. The sails could be mostly empty space to almost entirely empty space to reflect long wave radio-frequency phased array beams.

As for concerns about over burdening the conductive or super-conductive wires or cables used for such sails by extremely intense rF beams, note that such reflective members could be very conductive to superconductive to thereby yield near perfect reflection. The EM energy that was not reflected would largely pass through the openings in the sail grid.

Second, a magnetic and/or electric field based scoop or anti-scoop could divert the chargons away from the sail just as an extended electrodynamic scoop for an interstellar ramjet would. Electro-dynamic-hydro-dynamic-plasma-drive features could utilize the diverted plasma in a reactive and gainful manner.

The sail might be deployed in a manner that is orthogonal to the ship's velocity vector. The sail might be parallel to the spacecraft velocity vector and driven obliquely from behind. This way, the effective thickness of the sail could be thousands of miles and the sail could include electro-dynamic-hydro-dynamic-plasma-drive features.

Fourth, the above parallel sail could conceivably be made of negative refraction index materials that would be pulled forward by incident star light and highly blue-shifted CMBR, far infrared, and non-CMBR radio sources.

Fifth, the sail may be a deployed mag-sail or M2P2 type of sail or any other magnetic or plasma bottle sail. It is possible that a plasma affixed to the spacecraft could be driven by rf radiation. Upon attainment of extreme spacecraft gamma factors, laser light could be easily reflected by such sails. Plasma makes an excellent rF reflector even for very small plasma densities.

Sixth, some sail materials such as any future forms of super-strong very conductive to super-conductive metallic hydrogen can be used as nuclear fusion fuel for fusion rockets upon degradation to useless levels.

The general idea for obliquely oriented beams involves the beamed energy incident on both sides of the sail. The sail could include a surface of hair like cilia or any other surface contour that would work so as to much more effectively grab a hold of the light.

In addition, the sail could be fabricated from photovoltaic materials in order to provide power for electro-dynamic-hydrodynamic-plasma-drives or chargon rockets, or perhaps even photon rockets.

The sail might not need to be held by guy lines. A strong magnetic field based coupling or electrical charged based connection might work.

Another option is to fabricate the sail guy lines out of graphene, carbon nanotubes, boron nitride nanotubes, graphene oxide paper, and the like. A cable constructed from such materials could stretch for about 20 to 50 kilometers yet still handle tens to hundreds of Earth G's. The tensile strength of graphene is close to 18 million PSI for perfect forms.

The collection area of the sail can be very, very, large. A large electro-dynamic scoop could extent very far out from the sail.

Regarding nanotech self-assembly mechanisms, just simply greatly increase the capture area of a electrodynamic scoop to collect enough interstellar materials and use most of the collected interstellar material as an EHPD, an MHPD, or a combination of the two and use the rest of the materials for sail repair.

Regarding holding M2P2 plasma affixed to the ship under high gamma factor condition, simply increase the strength of the fastening fields.

Consider the interstellar matter density near our solar system of one particle for every 10 cm3. This density works out to be a layer of hydrogen or helium atoms about one atom thick for a column that is one light-year long: Not a show stopper for light sails or sails that are electro-dynamically shielded or protected!

If extreme materials are used with excellent reflectance, we could simply use a sail that has a thickness of one millimeter or more and which is monolithic, or better yet, use a sail with grid lines that are one millimeter or perhaps much greater in thickness. This way, a sail that has an area of only one square kilometer can intercept a beam having an equivalent black body temperature of several thousand Kelvins provided it is constructed of suitably refractive materials.

We could simply use electrodynamic methods of grabbing a hold of the interstellar gas and diverting around the spacecraft and sail. The power to operate the electrodynamic mechanisms can be supplied by beams. The electrodynamic methods can include lasers for ionization, or radio-frequency radiation where the gamma factors are suitably large, magnetic fields, electric fields, plasma fields affixed to the spacecraft, and the like.

Anyhow magnetic sails can be made of any ordinary conducting or superconducting period table materials.

It is also conceivable that a hybrid sail can be used where a current carrying magsail would deflect plasma away from a monolithic and grid-like light sail or radio-frequency sail.

Perhaps, there is no reason to worry about sail erosion in spite of the presence of interstellar gas for the following reasons.

First, extremely relativistic particles would likely deposit only a small portion of its energy within the sail thereby greatly lessening the number of atoms that would be knocked loose. This fact would apply to chargons as well as neutral incident particles.

Second, for sails of near micron thickness, atoms that were knocked loose would likely simply be re-assimilated by the bulk sail materials. Perhaps the only chance for an atom to be knocked loose would include atoms located on the backward side of the sail. Atoms for which its bonds where broken within the bulk sail material would tend to rebond with adjacent atoms.

Third, since the incident gas or plasma particle would deposit only a small portion of its energy within the sail, the kinetic energy per particle for particles that are knocked loose may be only slightly in excess of the binding energy of the dislodged atoms. Basically, the kinetic energy of the dislodged atoms could likely be re-absorbed and/or radiated away thereby promoting rebinding of the dislodged atoms.

Fourth, for cases where the sail would completely absorb the kinetic energy of the incident gas or plasma particles such as an alpha particle, for the case of a one micron thick sail, the sail would obviously be able to completely stop the chargon without losing it. Thus, any atoms disbonded by the incident chargon would also likely be captured and prevented from leaving the sail material.

Note that alpha particles from alpha decay processes of radioactive materials generally will pass through only a few microns of human flesh. The lethal effects of alpha emitters arise mainly from ingesting alpha emitter contaminants and alpha radiation short range destructive effects on cells especially DNA molecules. However, swelling of thin-film sails needs to be studied as alpha particles can build-up in sufficiently thick sails to cause the swelling.

Fifth, for grid-like sails, the grid lines might be positively chargeable so that incident interstellar or intergalactic ions are pushed away from the grid lines and through the openings within the grid-like sails. The effect would be analogous to the Van der Waals force that keeps neutral atoms from being squeezed together to tightly.

Sixth, theoretical studies of nuclear warhead decoys as inflated balloons in space made of sufficiently thin reflective membranes suggest that a 1 kiloton nuclear explosive detonated at a distance of 100 meters from such a decoys would not destroy the space-based decoys. The reason is such that the decoys membranes would cool so quickly by radiative cooling that the balloons could not reach temperatures that would corrupt the decoy composition. Sufficiently thin balloons would be almost transparent to the incident ionizing radiation. Studies suggest a radiative cooling on the order of hundreds of billions of Kelvins per second. The author of this paper was unable to locate any reference material to support the above values provided. However, Essig provides the above assertations on good faith taken from recollections of sources indicating the above provided estimates.

Appendix 1

Prior Art in Staged Compression Gas Guns and Large Artillery

There are many precedents for large-barreled guns that were able to achieve a maximum chamber pressure of about 255 mega-newtons/m^2 and projectile accelerations of $\mathbf{a} = \mathbf{F}/m = 27{,}000$ m/s^2. Here, the reader can conservatively assume a projectile mass of 2,700 pounds = 1,225 kg and a maximum chamber pressure of 2.55×10^8 N. The propulsive force on the round is assumed to be a maximum of 3.31×10^7 N. Note that these values are those achieved during World War 2 and which were reliably practiced in dozens of units as fielded on the third-generation battleships of that era.

> 1.0 caliber, .50 caliber, and two .17 caliber two stage light gas guns are housed in the Remote Hypervelocity Test Laboratory. These guns use gunpowder and highly compressed hydrogen to accelerate projectiles at speeds up to 27,500 feet per second to simulate impacts of particles on spacecraft and satellite materials and components.
>
> Two Stage light gas guns use gunpowder, the first stage, and highly compressed hydrogen, the second stage, to accelerate projectiles at high velocities to simulate orbital debris impacts on spacecraft and satellite materials and components.

First Stage

> The first stage uses conventional smokeless gunpowder as its propellant and works the same way as firing a bullet from a gun. The second-stage propellant uses a highly compressible light gas such as hydrogen. Since light gases have very low molecular weights, they are easily compressed to the high pressures needed to efficiently launch projectiles at hypervelocity speeds. The gun's breech contains the first-stage powder charge, set off by an electronic igniter. The ignition provides the explosion that drives a piston forward down the pump tube, rapidly compressing the hydrogen gas that provides the second-stage launch power. The front face of the piston is a hollow cone that forms a gas seal as it compresses the hydrogen at a speed of approximately 2,500 feet per second. The back end of the piston uses an O-ring to seal the expanding gases from the powder charge.

Second Stage

> The taper-bored, high pressure (HP) section referred to as the high-pressure coupling halts the propelled piston at the end of its travel down the pump tube. In the HP section, rapid internal pressurization is followed by an extremely high level of impact caused by the halted piston. A petal-valve diaphragm retains the gas until it bursts, and upon rupture, the launch package containing the sabot and the projectile is accelerated by the rapidly expanding light gas down the barrel into the expansion tank where a stripper plate separates the projectile from the sabot. The sphere enters the target tank (with air removed to replicate the vacuum of space) and hits the test article. The ASME-rated target tank is designed to contain the gas from the second-stage and the eruption of shrapnel and debris from the projectile's impact on the test article." [4].

FIGURE 1. Looking down the 1.0 caliber two stage light gas gun from the open end of the breech toward the target tank just outside the building, 175 feet away. Projectiles achieve speeds of 23,000 feet per second in just 24 feet of barrel.

Credits: NASA WSTF

Last Updated: Aug. 6, 2017

Editor: Judy Corbett

The US Navy's Iowa (four ships) class battleships carried a main armament of nine 16"/50 caliber guns in three triple turrets. The previous North Carolina (two ships) and South Dakota (four ships) classes carried a very similar main battery of nine 16"/45 caliber guns. These 10 ships, completed between 1941 and 1944, comprised the USN's "third generation" battleships and all saw service in WW II.

The designation 16"/50 means a 16" diameter shell and a barrel 50 calibers long. That would be 16x50=800 inches, or a barrel 66.66 feet long. The 16"/45 gun fired the same shells from a slightly shorter barrel 60 feet long.

The barrel had 96 rifling grooves (shades of Marlin's micro-groove type rifling, which is typical of cannons). The twist rate was one turn in 25 calibers, or 1:400". The maximum service pressure was 18.5 tons psi, or 370,000 pounds psi (corrected 37,000 psi).

Shells of different weights were fired, weighing from approximately 1,900 to 2,700 pounds. The heaviest shell was the AP (armor piercing) projectile, which had a maximum range 42,345 yards from the 16"/50 gun, or 39,000 yards from the 16"/45 gun (about 22 miles). [5]

According to Jane's Fighting Ships (circa WW II), the muzzle velocity (MV) was up to 2800 fps with a 2100 pound shell and muzzle energy was 98,406 ft. tons. The rate of fire was about two rounds per minute. [6]

So there is precedent for large-barreled guns that are able to achieve a maximum chamber pressure of about 255 mega-newtons/m^2 and projectile accelerations of a = F/m = 27,000 m/s^2. Here, the author conservatively assumes a projectile mass of 2,700 pounds = 1,225 kg and a maximum chamber pressure of 2.55 x 10^8 N/m^2. The propulsive force on the round is assumed to be a maximum of 3.31 x 10^7 N. Note that these values are those achieved during World War II and which were reliably practiced in dozens of units as fielded on the third-generation battleships of that era.

Bibliography

1. Landis, Geoffrey. "Interstellar Flight by Particle Beam." *Acta Astronautica* 55 (2004): 931–934.

2. Matloff, Greg L. "The Interstellar Ramjet Acceleration Runway." *Journal of the British Interplanetary Society* 32 (1979): 219–220.

3. Nordley, G., and A. J. Crowl. "Mass Beam Propulsion, An Overview." *JBIS* 68 (2015): 153–166.

4 Corbett, Judy. "Two Stage Light Gas Guns. Site Tour/Remote Hypervelocity Test Laboratory." https://www.nasa.gov/centers/wstf/site_tour/remote_hypervelocity_test_laboratory/two_stage_light_gas_guns.html.

5. Hawks, Chuck. "U.S. Navy 16" Battleship Gun Facts Chuck Hawks." https://www.chuckhawks.com/16-50_gun_facts.html.

To order additional copies of this book, contact:
Xlibris
844-714-8691
www.Xlibris.com
Orders@Xlibris.com

ISBN: 978-1-6698-6463-9 (sc)
ISBN: 978-1-6698-6462-2 (e)

Print information available on the last page

Rev. date: 01/28/2023

Printed in the United States
by Baker & Taylor Publisher Services